# DIGITAL DARK AGE

## A CAUTIONARY TALE

**Written by**
Matthew Connell, Louise Preston & Helen Whitty

**Illustrated by**
Matt Huynh

*The quicker we are to record and store our lives in digital form, the bleaker the future of information looks ...*

John Huxley, 'The Digital Dark Age',
*The Sydney Morning Herald*, 23 September 2005.

**CONTENTS**

| | | | | |
|---|---|---|---|---|
| **RECORDS TECHNOLOGY** *a story* | | 04 | – | 39 |
| Storytelling | | 06 | – | 09 |
| Clay tablet | | 10 | – | 13 |
| Letters and photographs | | 14 | – | 17 |
| Phonograph | | 18 | – | 21 |
| Magnetic tape recorder | | 22 | – | 25 |
| CSIRAC | | 26 | – | 29 |
| Discs | | 30 | – | 33 |
| Personal computer | | 34 | – | 37 |
| The Cloud | | 38 | – | 39 |
| **KEEPING OUR STUFF SAFE** | | 40 | – | 43 |
| **RECORDS TECHNOLOGY** *the real thing* | | 44 | – | 48 |
| **FURTHER READING** | | 49 | | |
| **GLOSSARY** | | 50 | | |
| **REFERENCES** | | 51 | | |
| **ABOUT THE PUBLISHERS** | | 52 | | |

Even ugly babies will one day grow up and want to see their baby photographs. Yet by then the images, sounds and even writings collected by their parents may have already disappeared. Hardware and software become obsolete, media degrade and memories slip away.

This book illustrates some of the important technologies we have invented to capture our stories and enable us to fetch and carry their threads across space and time. How clever we have been at making and using new technologies at a faster and faster pace, right up to the data explosion of the Internet.

We live in the present and embrace new technologies, but without effective systems in place to store our memories is it possible that, as Mark Twain once wrote, in the future we won't recall 'any but the things that never happened'?[1]

# Storytelling

ABOUT 30,000 BCE UNTIL ...

Once upon a time the only way to carry ideas across time and space was through storytelling in drawings, the spoken word or song. Memories were oiled with poetic devices such as rhyme, rhythm and meter, giving us stories to be remembered, told and re-told. The Indian Ramayana and Aboriginal Dreaming Stories are living stories for living cultures.

# Storytelling
## CONTINUED...OR NOT

Through the act of storytelling, information and ideas are remembered, reinterpreted and repeated. Only some stories survive. New ideas and concepts can be difficult to grab hold of without the technology of the written or recorded word.

# Clay tablet

2500 BCE TO ABOUT 100 CE

**C**ity life means trading for food rather than growing your own, even 'back in the day' before the first millennia. Stories no longer carried all the information currency required and people had to keep track of the activities of daily life including purchases.

# Clay tablet

**2500 BCE TO ABOUT 100 CE**

**CONTINUED...
OR CRUMBLED**

'Total: five grass-fed sheep. Total: one lamb. Total: four male kids.'

The calculation documented on a terracotta tablet issued from Drehem to Alulu in Sumer, 2041 BCE. Unromantic, mercantile, brief yet reliable, clay tablets were the proof people required. While indecipherable to the uneducated, clay tablets could be carried and stored. Many still exist today.

# Letters and photographs

ABOUT 100 CE TO ...

Letters feel special. They are composed with a person in mind even if a meeting with that person is never to be. Letters flow between loved ones. Sometimes they are read by strangers. Letters accompanied by photographs carry our personal luggage, memories and reminiscences. Photographs are still precious – just ask anyone who is packing to flee a fire.

# Letters ~~~~~~~ graphs

ABOUT 100 CE TO ...
THE LETTER CONTINUED TO DELIVER...

**W**riting gave stories and ideas permanence. Printing meant that knowledge could be collected, copied and distributed not just by monks with quills but also by merchants in the marketplace and young children at school. Marks on paper, written or printed, have survived for centuries and still speak to us with ease. Paper can also disappear in a puff of smoke or tiny shreds, or disintegrate in a mouldy cupboard.

> Sir,
> I beg to inform
> that on Saturday
> will be given
> the occasion of the
> from the English

# Phonograph

1877 TO 20th C

A machine that could talk made a celebrity of its inventor Thomas Edison. The phonograph was revolutionary because it had the ability to record and play back the same sound. The phonograph recorded sound vibrations with a needle tracing a line on a spinning wax cylinder. The needle retraced that line to play back the sound.

# Phonograph

187  Phonograph continued to play…

New technologies led us from making a mark on paper to making a mark in wax; from having technology to capture information to needing technology that also retrieves information. In the case of the phonograph, if its cylinder is still intact it is possible to play a recording using simple materials – a needle, cone and balloon.

# Magnetic tape recorder

1927 TO ...

A many-headed machine that could store and read sound on magnetic wire, tape or paper covered in magnetic dust. One head could read, another could write and another erased the sound. The magnetic tape recorder converted a sound signal into an electrical signal, which was then converted to a magnetic signal. For the first time, music programs could be pre-recorded or recorded in multiple parts and then mixed and edited. Originally invented for sound, it became the main way we stored computer-generated data, for sound, video and even museum and archival records.

# Magnetic tape recorder

1927 TO

*Magnetic tape doesn't always continue...*

**M**agnetic tape allowed us to create, store, erase, and re-use data, but as the tape ages and sheds its magnetic dust the data is at risk of being lost. Magnetic tape evolved to hold words, sounds and images, but only the same type of machine that encoded the data can also retrieve it.

# CSIRAC

1949 TO 1964

In the 1940s a new type of information machine arrived. This machine was made of hardware that you could touch, software that you couldn't and an inbuilt memory. Information was represented by noughts and ones. CSIRAC was the first digital computer in Australia and one of the first in the world. It was built for processing scientific and radio astronomy data accurately, sometimes taking all night to do so. One programmer had it playing digital music as early as 1950.

# CSIRAC

## 1949 TO

*CSIRAC CONTINUED... TO REDEFINE*

**W**eighing in at seven tonnes, CSIRAC 'filled a room the size of a double garage. While it required enough electricity to power a suburban street, it had only a fraction of the brainpower of the cheapest modern electronic organiser.'[2] It may have been a behemoth destined for a museum but CSIRAC was one thousand times faster than the best calculator of its age.

# Discs

1971 TO ABOUT 2000

IBM flipped the first floppy into our pockets and called it a Memory Disc. Like the reel-to-reel only much smaller, it stored data on a thin circular disc, encased in plastic. Only readable when placed in a computer slot, floppy discs distributed software, transferred data, and backed up data on a personal computer.

# Discs

1971 DISCS DISCONTINUED...

First there was the 8 inch disc, followed by 5 ¼ inch, 3 ½ inch, one sided, two sided, double density, high density, CD-ROM, CD, DVD, SD card and memory stick. The rate of change in both computer hardware and software was so rapid we didn't pause to transfer precious information from one disc to another – our memories lost through technological obsolescence.

# Personal computer

MID 1970s TO ...

Our love of electronic gadgets intensified from analogue through to digital, leaving the cynics behind, including Thomas Watson, Chairman of IBM, who in 1943 famously stated: 'There is a world market for maybe five computers.'[3] The personal computer has become an integral part of who we are and what we do; its use has redefined our culture.

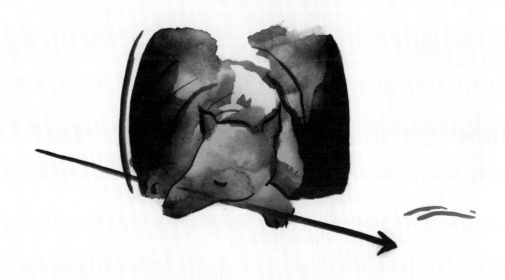

# Personal computer

MID 1970s TO ...

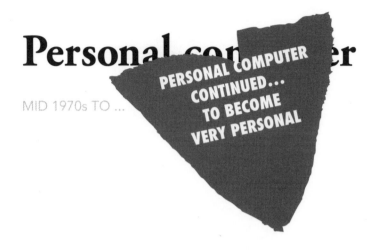

PERSONAL COMPUTER CONTINUED... TO BECOME VERY PERSONAL

The personal computer sits firmly between us and our photographs, songs, thoughts, ideas, social connections, contacts, school and work. From almost any location we can gather an abundance of information, take photographs, download music and video, and much more. From humble beginnings of just a few hundred bytes, memory space is now measured in gigabytes and terabytes; the sky's the limit when it comes to digital storage.

# The Cloud

1990s TO PRESENT

The Internet, a global interconnection of computer networks is becoming The Cloud. The Cloud can be a generous and giving space inviting people to share their stuff. We can deliver photographs to Flickr and personal information to Facebook.

The Cloud seems like a big library in the infinite sky, accessible even from a space station, but it is very much Earthbound. Cloud data is kept by large companies in refrigerated centres on server farms anywhere from North America to Pakistan with no friendly librarians to explain the cataloguing system.

Storytelling placed memories in a community where they were protected and re-told by a culture which valued and understood them.

The Cloud seems to be doing the same but is it enough when it comes to safely storing our memories and ideas?

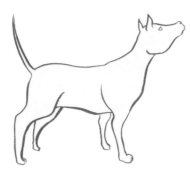

# KEEPING OUR STUFF SAFE
*Clicking 'Save' is just the start*

## 1

### KEEPING & FINDING IT

**Imagine if we kept all iPod or MP3 player files in one folder. No file names. Just sequential letters or numbers like a, b, c or 1, 2, 3. How would we find a song we want to hear?**

In libraries, museums and archives, where there are thousands of books, objects or records, numbers and letters form coding systems to uniquely identify each item. A **classification scheme** such as this can be implemented to group and organise data for better access to it.

Libraries may also use barcodes or Radio Frequency Identification (RFID) tags so that each item can be tracked using barcode or RFID readers. There are two components to the record about the item's movements – the code or tag and the machine required to read it. A complete picture of events is not possible without both the code and the **reading device**.

## 2

### DEFINING & MANAGING IT

**What if an iPod had two versions of a song by the same artist? Or there's a cover version and the original?**

Without control records to manage the songs it could take a while to find one. A program such as iTunes uses control records, tags or **metadata** to manage and describe song files, enabling better access to them. Artist, song title, album, year of release and genre of music are some examples. If we want to find a song to play quickly and cannot recall the artist, using this method we can look up the title or even just a word in the title. Applying this approach consistently and managing and organising our information well, results in having the data we need. Song found now ... and later.

# 3

## SECURING IT

**What if we decide to remove songs from an iPod and download new ones to it?**

We believe our music is safe because it is backed-up on computer or the original CD, so we move our new selection over. Then the computer crashes or the CD player breaks down and our music collection is potentially lost.

This scenario could apply to any collection of digital files. Organisations need to ensure their data is secure. For example, they need to pay their employees and this can only be done with **authentic** records that have **integrity** and are kept up-to-date and accessible. Pay transactions are generally in digital format; we are emailed a payslip, funds are electronically transferred to our account and we use a personal account card to withdraw cash from an automatic teller machine (ATM). If one of these processes goes wrong our finances can be seriously affected. Electronic records are especially vulnerable to damage or loss.

Businesses which run smoothly **back up** their computerised records, they keep copies or earlier versions with **record-keeping software** to enable restoration of lost data. We can take care of our own records by also keeping copies on separate digital storage devices.

# 4

## MOVING IT

**What if we transfer our songs to another computer and then find the old stuff won't play or songs disappear? Or suppose our teacher loses an assignment or essay? Or we can't find a favourite photo anymore?**

Technology, like living things, 'evolves'. The phenomenon of technological **obsolescence**, whether of hardware, software or due to degrading media is not new but it does pose a bigger threat as we are becoming increasingly dependent upon it.

Upgrading to the latest phone, for example, is exciting until we discover that the new phone is not compatible with the old one and we can no longer view the contact list we kept on it. Computer hardware and software upgrades happen at a rapid rate and can result in loss of compatibility with previous versions, sometimes preventing access to our own data.

**Conversion** software can be a remedy but it means more software! Software is often proprietary; there are a multitude of formats, which companies tend to update so they can sell new and improved versions. And what about our web browser, is it compatible?

# 5

## MOVING IT (CONTINUED)

One way clear is to use **'open', stable, standard file format** software. The specifications are available and have set **standards** so they can function across diverse computer information systems for long periods of time. We can also move our personal records to new systems through **migration/conversion**. Or we can **emulate** old software or hardware systems in a new system or **encapsulate** data to be viewed later. Another technique is **normalisation**, which extracts the content and essential characteristics of a record into a stable, accessible format.

## RELYING ON & SHELTERING IT

**Does our personal data have privacy and security?**

Some data in The Cloud, such as on Facebook, Gmail or YouTube, is owned by the organisations running those sites. We may no longer want to keep it all but this decision is not always ours. Some organisations have been known to disappear, taking our data with them. Hackers or viruses can also access our personal details if a site is not secure. It helps to be aware of the risks and to secure information with Personal Identification Numbers (PINs) and complex passwords, using only reputable websites, all the while remembering to **back up**, keep copies and recall our passwords.

# 6

## PROVING IT

**Imagine this scenario: our governing politicians promise to reduce a tax on chocolate. Time passes and not only do they break the promise, they increase the tax.**

If a year or two has passed, how will we remember? Can we rely on our individual memories to recall the detail of their promises?

Video footage and sound recordings provide evidence of politicians' statements and promises to hold them to account. Society's records provide proof of our authorities' activities, evidence of who did what, when, where, how and why. Even our own emails, text messages and tweets are records of our thoughts, our relationships, of who we are and what we do. Just like the humble, hand-written letter sent by 'snail mail'.

# 7

## DIGGING FOR IT ... MUCH LATER

**How many electronic devices do we rely on? Computers, iPods, iPads, mobile phones, e-books and more. How will our descendants access all of this data when we are gone?**

Our digital data lives inside these machines, which require a working device and a range of software to play back or use that information. Software and hardware are updated so fast that maintaining accessibility to data and ensuring its compatibility with old systems over periods as long as 5 to 10 years is a challenge. We need to think now about how we will store our valuable digital records for as long as we want to keep them.

By using open standards where possible, properly classifying information and moving it from old devices to new, we can ensure continued access to our precious data.

Technological progress is fantastic but we must also use it wisely so our information remains accessible. Whether communicated through spoken stories, clay tablets, paper or electronic documents, preserving our memories is a universal human desire. As long as we also remember that technology marches on!

# RECORDS TECHNOLOGY:
## *The real thing*

>>

*All objects are from the collection of the Powerhouse Museum, Sydney, unless otherwise indicated.*

### CLAY TABLET

Sumerian receipt for livestock written in cuneiform script on a terracotta tablet, 2041 BCE. 'Reciept issued - Total: five grass-fed sheep. Total: one lamb. Total: four male kids.' Purchased 1985. 85/452

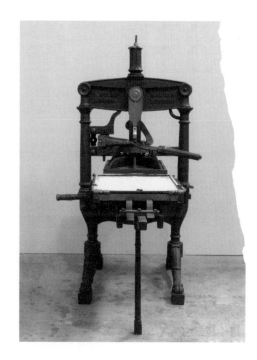

**LETTER**

This handwritten letter was sent from Tumut in country NSW to an engineer working at the Byrnes Flour Mill, Parramatta, in 1843. It describes the harsh economic times experienced by ordinary Australians at the time. From the Parramatta City Council Community Archive. ACC118/6

**PRINTING PRESS**

This Albion Press was used by Henry Parkes to publish the first editions of the Empire Newspaper during the early 1850s in Sydney. Albions were a hand-operated press in use from 1820 until the 1930s. Gift of Armidale Newspapers Ltd, 1929. H3408

RECORDS TECHNOLOGY:
*The real thing*

**PIONEER FAMILY PHOTO**

Pioneer settler Lucy Sawtell with her children in front of a cottage in Dorrigo, about 1895. Photographed by George Bell and published by Kerry and Co. From the Kerry Collection. 85/1284-777

**PHONOGRAPH**

Phonograph and wax cylinder, part of an Edison Standard Phonograph made in Germany by the Excelsior Company in 1900. Model manufactured under the Patents of Thomas A. Edison from 31 July 1888. Purchased 1971. H5298

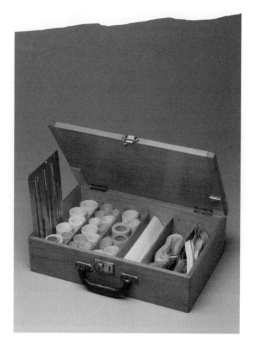

**MAGNETIC TAPE**

Half inch magnetic tape used for computer data storage. These tapes have been a mainstay of the computer industry since the mid 1960s. Teaching prop used by Powerhouse Museum staff.

**CSIRAC CARRYING CASE**

Carrying case with 'editing suite' used by CSIRO scientists working with CSIRAC from the late 1950s to 1964. It contains hand-editing equipment, programs and subroutine on 12-hole and 5-hole paper tape. Gift of John Spencer. 99/39/1

**RECORDS TECHNOLOGY:**
*The real thing*

### DISCS

A selection of floppy discs representing the most common standard sizes used between the 1970s and the early 2000s. The 3 ½ and 5 ¼ inch discs are from the curator's shelves and the 8 inch disc is from the Australian Computer Museum Society.

### ALTAIR 8800

Altair 8800 personal computer released in 1975 as an electronics kit by the MITS company in the USA. Often regarded as the first personal computer. 99/48/1

# FURTHER READING

## IN PRINT

Bettington, J., Eberhard, K., Loo, R. & Smith, C. *Keeping Archives 3rd Edition*, Australian Society of Archivists Incorporated, ACT, 2008

Ellis, J. A. (ed.) *Selected Essays in Electronic Recordkeeping in Australia*, Australian Society of Archivists Incorporated, ACT, 2000

## ONLINE

### www.powerhousemuseum.com/digitaldarkage

#### www.wepreserve.eu

Website for the following partners: DigitalPreservationEurope (DPE), Cultural, Artistic and Scientific knowledge for Preservation, Access and Retrieval (CASPAR), and Preservation and Long-term Access through NETworked Services (PLANETS)

#### xena.sourceforge.net

Website for Xena (Xml Electronic normalising for Archives). Developed by the National Archives of Australia, Xena is free open source normalisation software.

### www.howstuffworks.com

Website includes stuff on high-tech gadgets

### www.digitalpreservation.gov/videos/students10/index.html

Digital Natives Explore Digital Preservation

Digiman animations on YouTube (on 'wepreserve' account):

- Team Digital Preservation and the Deadly Cryptic Conundrum
- Team Digital Preservation and Nuclear Disaster
- Team Digital Preservation and the Arctic Mountain Adventure
- Team Digital Preservation and the Aeroplane Disaster
- Team Digital Preservation and the Metafor Common Information Model
- Team Digital Preservation and the Planets Testbed

Computer history film on YouTube (on 'ComputerHistory' account):

- Digital Dark Age – Revolution Preview

# GLOSSARY:
## What was that again?

**Archive:** place where records of continuing value, such as legal, evidential, financial or historical significance, are kept

**Authentic records:** those that document the facts they cover with proven reliability

**Back up:** copies made and checked regularly to ensure data is saved appropriately

**Classification scheme:** a pre-designated filing system

**Conversion (to paper or microfilm):** printing out to paper or copying to microfilm/fiche

**CSIRAC:** Council for Scientific and Industrial Research Automatic Computer

**Emulate:** when one computer imitates the hardware or software environment of another computer system, enabling data to be read

**Encapsulate:** to package information about records and their metadata in a way that ensures they can be read at a later date

**Integrity** where all parts of a record are obviously and logically connected to each other

**Long-term formats:** archival data formats such as ODF, PDF, PDF/A, XML, based on open standards (see 'open formats')

**Metadata:** information that describes or is associated with other data. Adds meaning by showing context, content and the structure of records, but on its own does not tell the complete story

**Migration / Conversion:** changing from one system or format to another. This involves moving data which may in turn be altered

# REFERENCES

**Normalisation:** migration of data formats to open format (e.g. XENA – XML Electronic Normalising of Archives)

**Obsolescence:** when old software, hardware and media deteriorate, degrade or are superseded as upgrades and innovations occur

**Open, stable standard file formats:** widely-used formats which apply formatting standards. Code is published and available enabling restoration of data

**Reading device:** the hardware, such as a computer, required to read data

**Record-keeping software or systems:** systems which control and identify records so they remain authentic, using metadata, access controls, audit trails and version controls

**Suitable environmental conditions:** for digital storage are those with a cooler temperature, lower relative humidity and protect against insects, dust and smoke

1. Paine, Albert Bigelow (editor) (1924) *Mark Twain's autobiography: in two volumes* New York, Harper & Bros., p 96

2. http://csiropedia.csiro.au/display/CSIROpedia/CSIRAC+-+Australia%27s+first+computer

3. Thomas Watson, chairman of IBM, 1943

First published 2011

**Powerhouse Publishing, Sydney**
PO Box K346 Haymarket NSW 1238
*and*
**Parramatta City Council**
PO Box 32 Parramatta NSW 2124

**Digital Dark Age: A Cautionary Tale**
Concept and text © Powerhouse Museum and Parramatta City Council
Illustrations © Matt Huynh

**Authors:** Matthew Connell, Louise Preston and Helen Whitty
**Project Editor:** Nicole Bearman
**Illustration:** Matt Huynh
**Publication Design:** Wil Loeng
**Printing:** Playbill
**Project team:** Nicole Bearman, Matthew Connell, Tracy Goulding, Helen Whitty (Powerhouse Museum), Louise Preston (Parramatta Heritage Centre) and Rebecca Pinchin (Regional Services, Powerhouse Museum)

Special thanks to Thomas Petr

This publication is copyright. Apart from fair dealing for the purposes of research, study, criticism or review, or as otherwise permitted under the Copyright Act, no part may be reproduced by any process without written permission.

**ISBN: 978 1 86317 136 6**

www.powerhousemuseum.com/digitaldarkage
www.parracity.nsw.gov.au/play/facilities/heritage_centre

## ABOUT THE PARTNERS

Museums, archives and libraries are the memory houses of our society. They collect, record, store, preserve and display our cultural heritage – the ideas and things valued by each generation. The technologies and systems for storing information are of particular interest to these institutions, as they believe that our understanding of the world is dependent on keeping information in or on some form of media.

**Powerhouse Museum**
Is part of the Museum of Applied Arts & Sciences, a NSW state government institution. The Museum conceives exhibitions and programs around the primary theme of 'human ingenuity'. Our programs are based on the ideas and technologies that have changed our world, and the stories of the people who create and inspire them. The Museum's huge and important collection might be perceived as meaningless unless it can be connected to ideas, and the community.

**Parramatta Heritage Centre**
Is part of Parramatta City Council and is co-located with Council's Visitor Information Centre. The Parramatta Heritage and Visitor Information Centre's passion is providing opportunities for people to engage with and understand the cultural heritage of Parramatta. Exploring issues surrounding access to and preserving that cultural heritage for future generations is a key part of its work. The Centre is enthusiastic about getting people to engage with discussions around these issues and to come up with innovative solutions. Its Archive department holds both Council's archival business records and a community archive containing historic records from past Parramatta people and organisations.

**Regional Services program**
This project has been supported by the Regional Services program at the Powerhouse Museum. The Regional Services program provides opportunities for partnerships between the Museum and organisations in Western Sydney and regional New South Wales, which support the documentation, preservation and display of the cultural collections of New South Wales.